Scientists of
Rivers, Lakes, and Ponds

By Eric Braun and Sandra Donovan

Raintree Steck-Vaughn Publishers
A Harcourt Company

Austin · New York
www.steck-vaughn.com

Copyright © 2002, Steck-Vaughn Company

All rights reserved. No part of this book may be reproduced or utilized in any form or by any means, electronic or mechanical, including photocopying, recording, or by any information storage and retrieval system, without permission in writing from the publisher. Inquiries should be addressed to Copyright Permissions, Steck-Vaughn Company, P.O. Box 26015, Austin, TX 78755.

Published by Raintree Steck-Vaughn Publishers,
an imprint of Steck-Vaughn Company.

**Library of Congress Cataloging-in-Publication Data
is available upon request.**
ISBN: 0-7398-4755-4

Printed and bound in the United States of America
1 2 3 4 5 6 7 8 9 10 WZ 05 04 03 02 01

Produced by Compass Books

Photo Acknowledgments
Corbis, 6, 34
Digital Stock, 10, 40 (top and bottom), 41 (bottom)
Jim Almendinger, title page, 22, 25, 26, 44
Photophile/Tom Tracy, 9; Sal Maimone, 32
Richard Axler, 30
Visuals Unlimited/John Sohlden, cover, 18; Tom Uhlman, 14; Inga Spence, 21;
   D. Cavagnaro, 28;
William Schlesinger, 16

Content Consultants
Maria Kent Rowell
Science Consultant, Sebastopol, California

David Larwa
National Science Education Consultant
Educational Training Services, Brighton, Michigan

This book supports the National Science Standards.

Contents

What Is a River, Lake, and Pond Biome? 5

A Scientist Across the United States 17

A Scientist in Minnesota's Lakes and Ponds . . . 23

A Scientist in the Rivers of the United States . . . 29

The Future of Rivers, Lakes, and Ponds 35

Quick Facts 40

Glossary 43

Internet Sites 45

Useful Addresses 46

Books to Read 47

Index 48

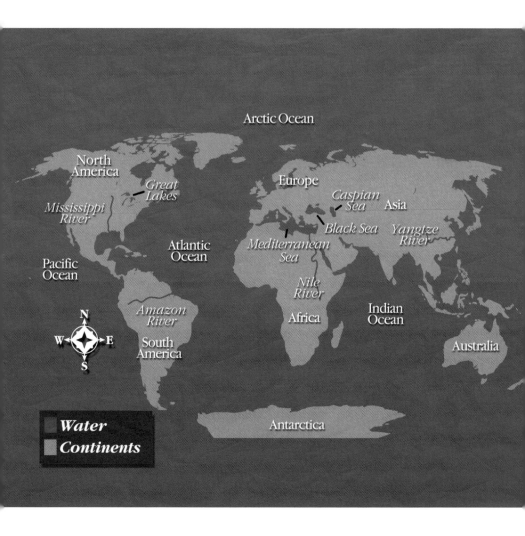

This map shows the location of the major waterways of the world, including several important lakes and rivers.

What Is a River, Lake, and Pond Biome?

Rivers, lakes, and ponds are a **biome**. A biome is a large region, or area, made of communities. A community is a group of certain plants and animals that lives in the same place. Communities in the same biome are alike in some ways. In the river, lake, and pond biome, for example, all the plants and animals live in or around water.

Some scientists believe that life began in the water billions of years ago. The very first plants and animals lived there. Water now covers about 3/4 of Earth's surface.

These are salt formations growing in the Dead Sea.

Saltwater and Freshwater

The two kinds of water on Earth are saltwater and freshwater. Oceans are saltwater. They make up most of Earth's water. Most rivers, lakes, and ponds are freshwater. Only 3% of all the water on Earth is freshwater.

Ponds and lakes form when water fills up holes in the ground. This water is still, meaning it does not move on its own. Ponds and lakes are found all over the world. Lakes are bigger and deeper than ponds, and they are also older.

Rivers are bodies of water that move in one direction. The place where they begin is called the **headwaters**. Often the headwaters come from lakes or melting snow in mountains. The end of a river is called the mouth. Rivers are found all over Earth.

Did you know that saltwater lakes can be as salty or even saltier than the ocean? The saltiest lake in the world is the Dead Sea in the Middle East. It is 10 times saltier than the ocean. The wind cannot make waves, and you cannot sink in the water because of the salt in it.

What Are Freshwater Biomes Like?

Some ponds are just a few square feet or meters in size. Some lakes are thousands of square miles or kilometers in size. Many ponds only last for a couple months. Lakes can last hundreds of years. In cold **climates**, small ponds can freeze solid in the winter. Most lakes are big enough that they do not freeze solid in cold climates.

The water in lakes and ponds changes temperature depending on how deep the water is. The water is warmest near the surface. This is because the sun's heat can warm it more easily. Most plants and animals live in this top area. The middle area of a lake is slightly darker and cooler. That is because less of the sun's light and heat reach all the way to the bottom. The deepest parts of ponds and lakes are cold and dark. Very little sunlight reaches these parts.

Rivers change from the headwaters to the mouth. The water is cooler at the headwaters than it is at the mouth. The water is also clearer at the headwaters. The mouth of the river

▲ This ship is sailing in a lake that has a frozen top layer with liquid water underneath.

usually flows into an ocean or a large lake. The water at the mouth has a lot of **sediment** in it. Sediment is rocks, sand, or dirt from the river bottom that mixes with the water. Sediment makes it hard for sunlight to pass through the water at the mouth.

▲ This flock of swans is part of the freshwater biome because swans live and hunt near water.

What Lives in Rivers, Lakes, and Ponds?

Plants and animals in rivers, lakes, and ponds have adapted to live in or around water. To be adapted means that something is a good fit for where it lives. Many plants and animals that live in rivers, lakes, and ponds could not live in another biome.

Different kinds of life are found in different parts of lakes and ponds. Plants live only in areas where sunlight can reach them. They need sunlight for energy to help them grow. Most plants grow in the water closest to the surface. Plants release **oxygen** into the water. Animals need oxygen to live. That is why most animals in lakes and ponds live near plants.

The middle and bottom depths of a lake receive less sunlight. The only plants that can grow there are **plankton**. Plankton are tiny plants or animals that drift or float. Some animals are able to live at the bottom of a lake. They eat plants or dead animals that sink from above.

Life in rivers changes as the river goes from the headwaters to the mouth. At the headwaters, some freshwater fish, such as trout, can be found. In the middle, more kinds of plants and animals are found. At the mouth of the river, there are fewer kinds of life. This is because the water is thick with sediment, so most plants cannot survive. Catfish and carp live in this part of the river. They have adapted to the sediment-heavy water.

Why Are Rivers, Lakes, and Ponds in Danger?

Water is important to all life on Earth. Some animals live in water. Almost all animals need to drink freshwater from rivers, lakes, and ponds in order to survive. People need water to drink. Some farmers spread water on crops to help them grow. People catch fish from rivers, lakes, and ponds. A lot of people also like to play in and live near rivers, lakes, and ponds.

Global warming can affect rivers, lakes, and ponds. Global warming is a slow but measurable rise in temperatures across all of Earth. Changes in temperatures, even by a few degrees, can cause changes in weather patterns. These changes mean some rivers, lakes, and ponds might receive more or less rain. Some might have higher or lower temperatures. New weather patterns may affect some plants and animals in rivers, lakes, and ponds.

Ponds can dry up when the weather gets too hot. Lakes, ponds, and rivers can overflow if it rains too much. These things can affect plants, animals, and even people.

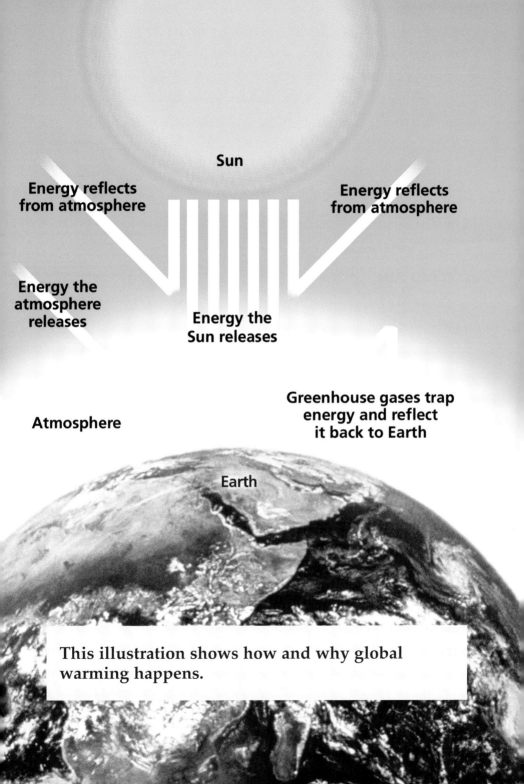

This illustration shows how and why global warming happens.

These people are helping to keep rivers, lakes, and ponds clean by picking up trash along the shore.

Pollution

People are also hurting river, lake, and pond habitats. A habitat is a place where an animal or plant usually lives. Many rivers, lakes, and ponds are **polluted**. Polluted means an area has been made dirty, especially with garbage. Some rivers, lakes, and ponds are polluted by **sewage**. Sewage is liquid and solid waste carried away in sewers and drains. Waste can be undigested food that leaves a person's or an animal's body in droppings. It can also be garbage, or something left over and not needed, such as chemical waste.

Rivers, lakes, and ponds also get polluted from **fertilizers** and **pesticides**. Fertilizers are materials used to make crops grow. Pesticides are chemicals used to kill pests, such as insects. Rain washes pesticides and fertilizers into rivers, lakes, and ponds.

Polluted rivers, lakes, and ponds can become **eutrophic**. Eutrophic means the oxygen in the water is used up faster than it can be replaced. Without oxygen, everything in the water can die.

Dr. William Schlesinger is taking measurements at a dry seasonal lake in New Mexico.

A Scientist Across the United States

Dr. William Schlesinger studies rivers, lakes, and ponds all across the United States. He is a professor at Duke University in North Carolina. He studies what the people, animals, plants, air, soil, and water on Earth are made of. He has worked in swamps in Georgia and deserts in California. He has also worked in forests in New England and North Carolina.

Schlesinger works in the desert in New Mexico, too. In the summer, it is over 100° F (37.8° C) almost every day there. Sometimes it does not rain for months. When it does rain, it can rain very hard. This rain can fill up lakes that are normally empty. Scientists call these special desert lakes "seasonal lakes."

These scientists are using a tool to gather river water to examine for pollution.

What Does Schlesinger Do?

Schlesinger uses science and tools to detect pollution in water. To detect something means to discover it. Some kinds of pollution are hard to detect in water. When you hold a glass of water, it might look like clean water. It could still be polluted, however. One kind of pollution makes too much **algae** grow. Algae are small plants or plant-like life without roots or stems that grow in wet places. Too much dead algae in a river will use up the river's oxygen. Without oxygen, the fish will die.

Schlesinger says that many of the things that people do change the quality of the air and water on Earth. These things include driving cars, growing crops, and building cities. Schlesinger uses special tools to measure these changes. Schlesinger can tell which changes are caused by people and which are caused by nature.

He says it is important to find out what changes are caused by things people do. If they know this, people can try to do these things less often. This will help preserve rivers, lakes, and ponds. Schlesinger says that people should try not to use fertilizers and pesticides that are bad for water. They should also try not to dump their waste into rivers, lakes, and ponds. These things can help to make rivers, lakes, and ponds healthier.

Keeping Freshwater Healthy

Schlesinger has learned a lot about pollution that is hard to detect. He hopes his discoveries will show people that they cannot dump waste into rivers. "When we do, we can expect fish kills and other problems," he says.

Some people in the government are trying to help Schlesinger in North Carolina. They are trying to pass laws to protect the rivers. The laws will say that people cannot dump too much waste into rivers.

Schlesinger is concerned about protecting the world's freshwater. "Very little of the water on Earth is freshwater," he says. "We must protect the quality of water for both humans and the other animals on Earth." This is why Schlesinger tries to understand the reasons the rivers, lakes, and ponds are changing. "Then we can manage them better, so that they will be available for future generations." A generation is a group of people born during the same time period.

Scientists like Schlesinger hope to stop people and boats from polluting freshwater.

Jim Almendinger is taking samples of sediment from the bottom of this Minnesota lake.

A Scientist in Minnesota's Lakes and Ponds

Jim Almendinger is a scientist who works for the Science Museum of Minnesota. He has spent many years studying lakes and ponds. Almendinger says you can tell a lot about lakes and the land around them by looking at the sediment they contain.

Almendinger really wants to find out what the climate was like at different times in history. He studies lake sediment to do this. The deeper into the bottom of the lake you go to collect sediment, the farther back in time you can study. "If you dig deep, you can tell what the climate was like thousands of years ago. If you go even deeper, you can tell what the climate was like millions of years ago," Almendinger says.

What Does Almendinger Do?

Almendinger says his favorite part of being a scientist is the time he spends at the lakes. This is called field work. Almendinger does most of his field work in Minnesota. He has also traveled to Alaska, Florida, Canada, and Sweden. When he is doing field work, he may collect some of the lake sediment to study. This is called a sample. In the summer, it is warm and sunny and easy to collect samples.

In the winter, the lakes and ponds may be frozen. Then, Almendinger drills a hole through the ice to collect sediment from the bottom of the lake. After he collects samples, he may bring them back to his science lab for more studying.

People who work in nature, such as forest rangers, ask Almendinger to help them find out about problems they have with their lakes and ponds. Almendinger brings his special tools and studies the lakes and ponds to find the answers to their questions.

Scientists use this tool to gather sediment and water from rivers, lakes, and ponds to study.

▲ Almendinger has done a lot of work to clean up lakes in Minnesota.

Keeping Freshwater Healthy

Almendinger's work has helped to clean up a lot of lakes around Minnesota. He also writes about his studies for magazines that are read all over the world. By reading his articles, scientists from many countries can learn from what Almendinger has learned.

Almendinger can tell how polluted the environment is from looking at lake sediment. Almendinger says pollution is the biggest danger to lakes and ponds. Plants and animals cannot live in polluted water.

Almendinger says scientists still have a lot to learn about lakes and ponds. "We will learn more about how we can live near lakes without damaging them," he says. He wants to do more to create clean lakes and ponds for the animals and plants that live there.

Almendinger once found a forest at the bottom of a lake? He was studying sediment at a lake in Alaska when he made his discovery. "A helicopter dropped us off on a raft," he says. "We were surrounded by forest, and looking down into the water we could see that the trees continued right down to the lake bottom." This strange forest did not grow under water. It grew first on some soil that was resting on top of a huge block of ice. Hundreds of years later, the ice melted. Then, the whole forest sunk down into the hole where the ice had been. The hole had become a lake from the melting ice.

Richard Axler studies rivers around the United States, such as the Mississippi River, shown here.

A Scientist in the Rivers of the United States

Richard Axler is a scientist who works at the University of Minnesota. His studies rivers, streams, lakes, and ponds across the United States. He says his favorite places to work are beautiful mountains, deserts, and forests when the weather is good. Many times the weather is not good and he has to work anyway. "When you're freezing cold and wet and tired and nothing is working, it's miserable," he says.

One time in Nevada, Axler and another scientist hiked 3 miles (4.8 km) up a desert stream. The temperature was 123° F (50.6° C). "It was so hot that we had to walk in the stream because the ground was too hot, even with tennis shoes," he says.

▲ Axler's assistants are taking samples of river water to study.

What Does Axler Do?

Axler says there are four important steps in science. First, he goes to the library to learn all he can about what he is studying. Next, he collects plants and animals from the river he wants to study. Then, he measures the plants and animals in different ways. Finally, he puts the whole story together and writes a report about what he has found.

Axler likes to study rivers and streams because water is so important to plants, animals, and people. He says that there is very little freshwater in the world, so it is important to find out how to keep it clean. "One interesting thing about studying rivers and streams," Axler says, "is that the water is always moving. The water you study at one place in the river is gone right away. Sometimes you have to follow the water as it moves down the river."

One time Axler got caught in a blizzard in northern California. A blizzard is a severe snowstorm. He was riding on snowmobiles with other scientists. "We had an adventure that ended up with two snowmobiles at the bottom of a lake," he says. The next summer, they went **scuba diving** in the lake and pulled out the two snowmobiles. "They still run!" he says.

Sometimes people change the flow of a river by building a dam like this one.

What Has Axler Found?

Axler has found many problems caused by people. One big problem is that pesticides and fertilizers are washed into rivers and lakes by rain. Another problem is mud and other soil that washes into rivers from farms and city streets and parking lots. Axler says that this soil has many things in it that harm plants and animals in rivers. It can be hard for fish to breathe with all these things floating in the water. Also, it can be hard for animals to find food when the water is very cloudy.

People can change the flow of rivers by building dams. They can also hurt rivers when they build new houses or buildings right along the shore. When people build there, they often cut down plants. Other plants and animals cannot live without these shore plants. Axler works to help people find out how to fix these problems. "Finding out what the problem is can be the most important step," he says.

These people are recycling. Recycling helps lessen the pollution of the freshwater biome.

The Future of Rivers, Lakes, and Ponds

The future of rivers, lakes, and ponds depends on people. With so little freshwater in the world, it is very important to take care of it. People cannot drink saltwater or water crops with it. Today, however, many freshwater biomes are polluted or are losing large amounts of water.

Polluted freshwater can be saved if people try to clean it up. People can help by recycling trash and using less water. Recycling means taking old things, such as empty cans, and using them to make new things. People can also save water by turning off the faucet while brushing their teeth and by fixing leaky faucets.

This map shows where the Great Lakes of North America are located.

Where Are Rivers, Lakes, and Ponds Protected the Best?

Rivers, lakes, and ponds are best protected in national parks. A national park is a park that is owned and protected by the government. People who work at national parks make sure that pollution does not get into the water. Rivers, lakes, and ponds are found in many different national parks people can visit, including Yosemite National Park and Yellowstone National Park.

Lake Erie, one of the five Great Lakes of North America, is a lake that was saved because people worked to clean it. It used to be badly polluted. The waste of all the people who lived around this huge lake ran into it. It looked like pollution was going to kill all life in Lake Erie. In the mid-1960s, however, the people who lived around it made laws to stop the dumping of waste. Today, Lake Erie is much cleaner.

The Endangered Alabama Cavefish

One of the endangered animals of the lakes is the Alabama cavefish. Endangered means the animal is in danger of becoming extinct. The Alabama cavefish is a white fish about 2 inches (5 cm) long. They cannot see at all and are found in only one cave, called the Key Cave, in a lake in Alabama.

About 20 years ago, scientists discovered that there were less than 100 Alabama cavefish living. Scientists think there are many reasons for this. One reason is that the water in the cave was becoming polluted. Waste, pesticides, and fertilizers were being washed into the lake around the cave.

In the last 20 years, scientists have done many things to help protect the Alabama cavefish. They built a fence near the lake to keep people from disturbing it. They are also trying to find out how to stop pollution in the area.

You can help by contacting the U.S. Fish and Wildlife Service. It is the job of people there to

This map shows where Key Cave, the only home of the Alabama cavefish, is located.

teach about the Alabama cavefish and other endangered animals. You can write them to ask for more information about what you can do to help endangered species where you live. Then, you can teach your friends how to help, too.

Quick Facts

Millions of birds use the habitats found in rivers, lakes, and ponds for their nests and their food.

Lake Baikal in Siberia is the oldest and deepest lake in the world. It is at least 25 million years old. Many plants and animals that live there are found nowhere else.

The hippopotamus can hold its breath under the water in rivers, lakes, and ponds for up to five minutes.

When the Amazon River floods, it covers a forested area as large as England.

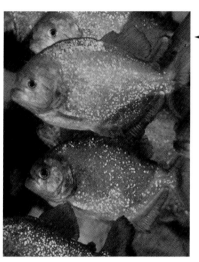

Piranhas are South American fish that eat fruit and nuts that fall into the water. When fruits and nuts are hard to find, piranhas eat animals that go into the water. Groups of piranhas can eat a whole animal, such as a goat, in only a few minutes.

The osprey is a bird of prey that lives in the Everglades. It eats mainly fish.

Beavers are mammals that build dams on streams and rivers, turning the area behind the dams into ponds.

On pages 45-47, you can find information that tells how to help save rivers, lakes, and ponds.

Glossary

algae (AL-jee)—plants or plant-like life without roots, stems, or leaves that lives in or under water

biome (BYE-ome)—large regions, or areas, in the world that have similar climates, soil, plants, and animals

climate (KLYE-mit)—the usual weather patterns in a place

eutrophic (yoo-TROH-fik)—a body of water rich in nutrients; the nutrients in eutrophic waters often cause all of the oxygen to be used up, so plants and animals can no longer live there

fertilizer (FUR-tuh-lize-ur)—a material put on land to make crops grow better

headwaters (HED-wah-turz)—the beginning point of a river

oxygen (OK-suh-juhn)—a colorless gas found in air that plants and animals need to breathe and fires needs to burn

pesticide (PESS-tuh-side)—chemicals made by people used to kill pests, such as insects

plankton (PLANGK-tuhn)—tiny plants and animals that float or drift in lakes and oceans

polluted (puh-LOOT-ed)—when an area has been made dirty, especially with garbage or other things made by people

scuba diving (SKOO-buh DIVE-ing)—underwater swimming with an air tank connected to the mouth by a hose

sediment (SED-uh-muhnt)— rocks, sand, or dirt in a body or water or other liquid

sewage (SOO-ij)—liquid and solid waste carried away in sewers and drains

Internet Sites

Biomes of the World
http://mbgnet.mobot.org/
Learn information about the different biomes around the world and where they are located.

Biomes/Habitats
http://www.allaboutnature.com/biomes
Find a description of each biome and information about the exciting animals that live there.

Lakes and Ponds
http://www.twingroves.district96.k12.il.us/
    Wetlands/LakesPonds/LakesPonds.html
Learn about the fish, birds, insects, and mammals that live in and around lakes and ponds.

Water on the Web
http//:wow.nrri.umn.edu
Explore how to use science to save the environment here.

Useful Addresses

National Wetlands Conservation Project
1800 North Kent Street
Suite 800
Arlington, VA 22209

National Wildlife Federation
1400 16th Street NW
Washington, DC 20036

Sierra Club
330 Pennsylvania Avenue SE
Washington, DC 20003

U.S. Fish and Wildlife Service
Publication Unit, Room 148
1717 H Street NW
Washington, DC 20240

Books to Read

Amos, William H. *Life in Ponds and Streams.* Washington, DC: National Geographic, 1981.
Learn about the animals and plants that live in ponds and streams.

Kosek, Jane Kelly. *What's Inside Lakes?* New York: Rosen, 1999.
Explore the living and nonliving things you can find inside lakes.

Parker, Steve. *Pond and River.* Eyewitness Books. New York: Alfred A. Knopf, 1988.
Compare ponds and rivers and their special features.

Stone, Lynn M. *Pond Life.* Chicago: Children's Press, 1983.
Discover all of the interesting plants and animals, such as frogs, that live in ponds.

Index

Alabama cavefish, 38
algae, 19
Almendinger, Jim, 23-24, 26, 27
Axler, Richard, 29-31, 33

biome, 5, 8, 10, 35

climate, 8, 23

eutrophic, 15

freshwater, 7, 12, 20, 31, 35

global warming, 12
Great Lakes, 37

headwaters, 7, 8, 11

oxygen, 11, 15, 19

plankton, 11
pollution (polluted), 15, 18, 20, 27, 35, 37, 38

recycling, 35

saltwater, 7, 35
sample, 24
Schlesinger, William, 17-20
seasonal lake, 17
sediment, 9, 11, 23, 24, 27